Dr Hairy's Guide to the GP Curriculum

By Dr David Hindmarsh

and Julian Le Saux

This is a bit out of date now — but hopefully still quite funny.

© 2007 by Dr David Hindmarsh and Julian Le Saux
ISBN 978-1-4092-1959-0

Self-published via Lulu (www.lulu.com)

All situations, incidents and attitudes in this book are entirely fictional. There isn't really a GP called Dr Hairy – or if there is, we've never met him –and no real-life GPs, registrars or patients would ever actually behave in the manner described in this book.

david.hindmarsh@nhs.net

julianlesaux@nhs.net

Contents

About the authors   5
Introduction   6
About this Guide   9
Other Resources   10
What should be taught and learned where?   11
The Six Domains   15
    Primary Care Management   15
    Person-Centred Care   24
    Problem Solving   27
    Comprehensive Approach   34
    Community Orientation   39
    Holistic Approach   42
The Three Essential Features   45
    Contextual Aspects   45
    Attitudinal Aspects   47
    Scientific Aspects   49
Core and Extended Statements   51
    Medicine   54
    The general practice consultation   58
    Ethics and Values Based Medicine   60
    Evidence-based health care   63
    Information Management & Technology   65
    Healthy people: promoting health and preventing disease   68
    Care of children and young people   70
    Gender-specific health issues   71
    Sexual health   73

Care of people with cancer and palliative care	74
Care of people with mental health problems	75
Care of people with learning disabilities	76
Drug using adults	77
ENT and facial problems	79
Skin problems	80

Appendix: A game   **81**

About the authors

Dr David Hindmarsh has been a GP in Cranbrook, Kent for 17 years. He is a Trainer and a Course Organiser. Despite this extremely rigorous and demanding way of life he is extremely kindly and patient and has never even lost his temper once, not ever. He attributes this extraordinary forbearance to the large number of biscuits he consumes every day.

Julian Le Saux is Dr Hindmarsh's practice manager, and he has been doing this job for 17 years, even though he's really got better things to do. Despite his extremely onerous and boring duties he manages to stay almost superhumanly cheerful and never has a cross word with anyone. He attributes this extraordinary forbearance to the large number of sarcastic remarks he makes every day, most of them directed at Dr Hindmarsh.

Introduction

The new curriculum for registrars (or the RCGP Core Curriculum, which is its proper title) is a deterring document, approximately 600 pages long. It is also organised in a way which, at first sight, may seem a little confusing. At second sight it may seem completely bewildering, and at fourth sight it will probably seem absolutely terrifying: but at fifth or sixth sight, with the help of this guide, it may start to make a bit of sense; so stick with it.

There are three different levels to the curriculum:

1. Firstly, there's a great big list of all the different subjects you're supposed to learn - mostly (but not entirely) clinical subjects. For example, you're supposed to know about

    - cardiovascular problems
    - care of people with mental health problems
    - genetics

    These are just examples. But you're also supposed to know about administrative and managerial matters such as

    - clinical governance
    - information management and technology.

    Again, these are just examples. This great big list is called "Core and Extended Curriculum Statements" (don't ask me why), and it forms, as it were, the bottom layer of the curriculum.

2. On top of this big list sits a second layer made up of six domains or core competencies –

   - Primary care management
   - Person-centred care
   - Specific problem-solving skills
   - A comprehensive approach
   - Community orientation
   - Holistic approach

3. On top of this again sits a top layer made up of three essential features –

   - Context
   - Attitude
   - Science

Now, the difficult bit to grasp is this: the three layers aren't supposed to be learnt separately from each other. When you learn all the stuff in the great big list of different subjects, you're supposed to learn it in the light of and in combination with the Six Domains and the Three Essential Features. Ultimately these three layers are supposed to combine together into one mystical and harmonious body of knowledge and understanding, realization and enlightenment, which constitutes the ideal learning and preparation of the ideal GP registrar. Good luck with that.

Still confused? Well, think of it like this. Imagine you were setting out to become a chef. You might start by learning all the different ingredients that are used in cooking: meat, fish, vegetables, fruit, herbs, spices and so forth. But if you went to a restaurant equipped with only this knowledge and asked for a job, the head chef would tell you that you didn't really know anything about cooking at all.

In order to be a cook you not only have to know your ingredients: you also have to know what to do with them: how to peel them, slice them, combine them and cook them. But again, if you learnt all this stuff and went back to the restaurant looking for a job, the head chef would tell you that your knowledge and skills still weren't enough by themselves.

In order to be a real cook, as distinct from someone who just knows how to follow a recipe, you need more than a lot of knowledge and a set of skills: you need an instinctive understanding of food. You need to respond to the way it looks, the way it smells and above all the way it tastes. And you need to be able to combine this instinctive stuff with the detailed knowledge and the skills and techniques, so that all three layers are working together at the same time when you're cooking.

In just the same way, good GPs don't only need a lot of detailed knowledge about coronary heart disease and so forth; they also need to acquire the skills and techniques with which to put this knowledge to good use; and on top of this they need an aptitude for the practice of primary care itself, a feeling for the way in which its scientific content plays out through its social context and the attitudes of the people involved. And they need to be able to put all these different things together into a single body of expertise when they are at work.

Now I've made myself really hungry and I'm off to eat a Mars Bar.

About this Guide

There are already a number of guides to the nRCGP Curriculum available (see below, Other Resources) and new ones are being brought out all the time. This is because lots of people (particularly the Deaneries, which are responsible for teaching it) have realised that people are going to need some kind of easily-digestible introduction to the Curriculum, because it's so big.

Most of these guides, however, are written in rather formal language, and they don't tend to give practical examples of what they mean. When you read something like "GPs have a responsibility for the community in which they work that extends beyond the consultation with the individual patient", you tend to end up with a rather blank mind. You can tell that it means something, but it's difficult to focus on exactly what it does mean. But if the next sentence reads "For example, if the patient has bird flu, it might be an idea to report the case to the local Public Health department", then things start to make a bit more sense.

This guide, therefore, attempts to make the principles of the Curriculum easier to understand by giving illustrations of what they might mean in practice. It also tries to use ordinary language wherever possible, just like those Websites for Dummies and Computers for Absolute Thickos books which have proved so popular in recent years. And just as a bonus, it's got some jokes.

What it doesn't do, unfortunately, is get you out of having to learn all the stuff that's in the Curriculum.

Other Resources

It's difficult to imagine why anyone would want to read anything else about the new Curriculum after reading this Guide: in fact some people may feel inclined to give up reading altogether. Some may even be tempted to give up medicine. However, there are other resources available which will be a big help to you in your training, so you really ought to check them out. The starting-point is the RCGP Curriculum Website, which is at http://www.rcgp-curriculum.org.uk/ . This has got all sorts of useful things on it, including the official Curriculum Documents (in downloadable form), a Student Forum, a Curriculum Map, and information about the e-Portfolio.

There's also a book called *The Condensed Curriculum Guide* by Riley, Haynes and Field (ISBN 9780850843163).

What should be taught and learned where?

The Curriculum puts a certain amount of emphasis on the apprenticeship model of learning, which suggests that most of your learning will come either in the form of explicit instruction from your trainer, or from observing and absorbing how your trainer does things. The problem with this is that your trainer is only human, and may not be a completely ideal role-model as a GP. If your trainer keeps falling asleep and snoring loudly during consultations, are you supposed to do the same? If your trainer keeps getting his foot stuck in the waste-paper bin under his desk, or his tie wedged in the shredder, are you expected to follow suit?

Clearly the apprenticeship model has its limitations, and it should be borne in mind that in adult learning, learners need to be self-motivated, self-aware, responsible for their own outcomes and in control of their own learning agendas. Your trainers will do their best to train you, but in the end it's up to you how much you get out of your time as a Registrar, and ultimately whether you pass or fail.

The model of development in the curriculum is actually the Dreyfus and Dreyfus model of becoming an expert. This model suggests that you will become an expert when you have

- Adequate experience
- outside recognition
- knowledge

Knowledge can be thought of as

- Propositional knowledge (facts)
- tacit knowledge (experiential, skills, intuition and pattern recognition)

"Experts" often make decisions based on pattern recognition and intuition (or gut feelings) and may find it hard to explain the process involved in these decisions. To put it another way, there's more to becoming an expert than learning a lot of facts and ticking a lot of boxes. You have to develop a "feel" for general practice.

While you're in training you will be expected to do your learning in a number of different ways:

- Working in General Practice
- Working in Hospital
- Half-day release course
- Independent training

Previously most trainees were only able to spend 12 months in general practice itself, and the rest of their time was spent going from speciality to speciality in the Hospital. Research has shown, however, that experience in general practice is valued more highly than hospital experience not only by current trainees, but by recently qualified GPs and by educators as well.

While working in primary care you will get your learning by:

- Observing established GPs and other primary care practitioners
- Supervised surgeries followed by unsupervised surgeries
- Joint surgeries with a GP trainer

- Reflection on learning
- Problem case analysis and random case analysis
- Etc, etc…

In other words you'll spend a certain amount of time sitting in with other GPs (especially your trainer), practice nurses, district nurses, the practice manager and so on, to see how things are done; then you'll have a go at doing things yourself; then you will need to reflect on what you have learned; and you will also need to talk things over and get feedback from your colleagues, especially from your trainer.

Although there's less emphasis on hospital learning than there used to be, it's still a very valuable part of your training. For one thing, hospital attachments provide exposure to higher numbers of seriously ill patients. For another thing, they give you insight into what happens on the other side of the primary care/secondary care fence, and this insight will be absolutely invaluable throughout your life as a GP.

The half-day release course was sometimes regarded as something of an "optional extra" for GP Registrars, partly because the training it offered was rather unstructured. Under the new curriculum, however, it is more structured and less optional. Here are some methods of teaching and learning used in the half-day course:

- Presentations
- Groupwork
- Role play
- Storytelling (scenarios from our practice)
- Significant event analysis

- Case-based discussion
- Proposition learning/theoretical - microteaching

As part of the half-day release course registrars are expected to research certain subjects and then teach them to their peers. This is an important aspect because teaching is often the best way to learn.

> *Zen Poem*
> When I look, I do not see
> When I listen, I do not hear
> When I teach, I learn

The half-day release course also gives you a chance to meet all the other Registrars in your group, have a good old gossip, and complain about all your trainers.

Throughout your training your progress will be assessed through a process of Workplace Based Assessment. You can find out more about this by downloading a publication called "A Brief Guide to Workplace Based Assessment" from the Royal College of General Practice – the web-address is http://www.rcgp.org.uk/docs/nMRCGP_WPBA_in_the_nMRCGP_sep07.doc (but it's no joke transcribing that into a browser, so it might be easier just to look for it on the RCGP website). An important element of the assessment process will be the e-portfolio, which allows you to build up an electronic record of your training, which can be shared and talked over with your trainer. More information about the e-portfolio is available from http://www.rcgp.org.uk/pdf/Trainees%20Guide.pdf .

The Six Domains

Domain 1 - Primary Care Management

Managing patient care pathways

In Primary Care, general practitioners work within an extended team. This means that there are lots of people (and institutions) all over the place, providing all sorts of services, and you're supposed to know something about them so that you can tell your patients which care pathways would be most appropriate to them.

- If a patient comes to you with an "unselected problem", you don't necessarily have to be able to make a diagnosis then and there: but you have to be able to decide how urgent the problem is, and which further investigations or care pathways are appropriate. It's no good sending your patient to the obstetrics department if he's an old man complaining of a pain in his big toe.
- It's also no good sending your patient to dermatology for a wart, if the latest dermatology protocol says they don't treat warts, they only treat "warty lesions" which look as if they might be malignant.
- Similarly, if your patient is a pregnant woman it's no good referring her to obstetrics at West Hospital if the obstetrics department has just been moved to East Hospital. You need to know who does what in the extended team; and if you're not sure, you need to know where you can find out.

- You're also not supposed to refer your pregnant patient to East Hospital if she has specifically stated that her sister went to East Hospital and said it was awful, so she doesn't want to go there. You should tell her what the alternatives are, and allow her to make an informed choice.
- You're meant to act as an advocate for the patient – not one of those strange green vegetables you find in Tescos, or one of those thick yellow drinks your grandma used to have at Christmas, but someone who speaks up on the patient's behalf - which means that if the hospital isn't offering an appointment until the month after next for an urgent condition which you think ought to be seen the day after tomorrow, you've got to do something about it (apart from sighing, rolling your eyes and saying "Typical NHS").

Managing collective practice

The above section is all about you, as an individual practitioner, knowing how to choose the best care pathways for your individual patients. But remember that when you work in the NHS you don't operate purely as an individual, even though your involvement with other professionals may not be as obvious as it would be in a hospital.

- You'll probably be working in a surgery alongside other GPs, and you will all have a collective responsibility for the care of all the patients on the practice list. This collective responsibility will work much more smoothly if you're all doing more-or-less the same things.
- The surgery will almost certainly be part of a Practice Based Commissioning "cluster", which means that a whole group of local

surgeries will be trying to work collectively in order to control referral rates and prescribing costs.
- Your surgery will also come within the catchment area of a PCT (Primary Care Trust), which will be running various protocols, enhanced services and goodness knows what else of which you ought to be aware.

What does this mean in practice? Well, if you start every patient whose fasting cholesterol is above 5.0 on statin treatment, whereas the other GPs in your surgery only use statins for patients with Framingham scores of 20% or higher, then

- The patients will get very confused, because one doctor in the practice will be telling them they ought to be on a statin, whereas the others will be telling them there's no need;
- Your GP colleagues will wonder what the hell you're playing at, and will eventually ask you this question out loud, even if they don't phrase it like that; and
- The pharmaceuticals officer from the Practice Based Commissioning group will come sniffing around to find out why the prescribing rate for statins in your practice has suddenly shot up.

Exercising judgement and making choices

Care within the NHS often involves something called variety of delivery, sometimes also known as plurality of delivery, which basically means that you can get the same service from several different places.

Where does your surgery get its milk from? Every surgery needs lots of milk, because everyone who works in a surgery spends all day drinking coffee and tea (not to mention eating biscuits, cakes and peanuts). This milk can come from

- The milkman
- The all-day shop across the High Street
- The supermarket, which is a bit further away

In other words, you can get the same thing from several different places. But each of these places has its advantages and disadvantages:

- The milkman delivers to your door, which is convenient; and the milk bottles get recycled, which is environmentally friendly; but it costs more, and when the milkman comes to collect the money he won't stop talking.
- The all-day shop is cheaper, but they tend to sell out of normal milk, which means that you end up having to buy semi-skimmed or UHT, and everybody moans.
- The supermarket is cheaper still, and they always have normal milk, but it's a right nuisance having to go up there, and they rip off the poor farmers.

What do you do under these circumstances? Well, you have to exercise your judgement. There are no clear-cut right or wrong answers: you have to weigh up the advantages and disadvantages. Or be like me, and drink your coffee black.

Let's look at a similar example in general practice. A patient comes to see you complaining of depression (if this hasn't happened to you yet, then you haven't been in general practice very long). Three options occur to you:

1. You could prescribe anti-depressants
2. One of the practice nurses sees patients for "problem-solving": you could refer the patient to her
3. The local hospital has a psychology and counselling service

As with the milk, each of these options has advantages and disadvantages.

1. A prescription for anti-depressants is the simplest thing to organise, and it doesn't take up any nursing time, but it costs the NHS money.
2. The nurse who does "problem-solving" can't deal with patients who are severely depressed, so you have to get the patient to fill out a depression questionnaire before you can refer to her.
3. The psychology and counselling service can deal with more severe or complicated cases, but they have a long waiting-list, and they are the most expensive of the three options.

Again, there may be wider management issues to consider, for example:

- The other GPs in your practice have adopted a depression management protocol, which involves using questionnaires on all depressed patients before treatment is initiated.
- The PBC group have issued an advice-sheet about cost-effective use of anti-depressants.
- The PCT are running a Local Enhanced Service for depression, which means that every patient diagnosed with mild to moderate depression but treated without referral to the hospital attracts a small fee.

Time management

You can't do anything properly unless you give yourself time. There now follows an extremely vulgar example.

Everybody needs to poo, and for this reason everybody sets aside a certain amount of time every day (or, in some hard-bitten cases, once every few days) during which he or she can sit in the bathroom alone, doing what needs to be done in an unhurried and thorough fashion.

But let's imagine that you scheduled your week so badly that you never had time to go to the toilet. Or let's imagine that you gave yourself just enough time to get to the bathroom, but not enough time to take down your undergarments, lift the lid or assume a seated position. The results would be unfortunate, wouldn't they?

This is the kind of situation into which people get themselves if they can't manage their time. They know that certain things have got to be done, but they don't set aside any time in which to do them: or they set aside enough time to make a start, but not enough time to do them properly.

GPs, for example, give themselves enough time to do their consultations and their visits, but not enough time to write them up properly. Or they don't allow sufficient time for writing referral letters, processing lab results or reading. Or they do all of the above, but don't allow themselves any time to meet with their colleagues, with the result that things never get talked over properly within the team.

Result? Crap everywhere.

Leadership, and managing change

The pressures on the NHS are building up all the time:

- People are living longer
- The birthrate is falling, so there are more old people with fewer young people to look after them
- "Affluent diseases" such as obesity, weight-induced diabetes and alcoholism are on the increase
- Patient expectations are on the increase too

As a result of these constantly-increasing pressures the NHS is never far from the headlines, because things keep going wrong with it. As a result of being always in the news, the health service is perceived as a big vote-winner or vote-loser, which makes it into a "political football". The Government and the Department of Health keep launching new initiatives, some of which are practical attempts to control things like waiting-lists and Harold Shipman, and some of which are merely vote-grabbing publicity-stunts.

It's a bit like one of those wooden ships during the Napoleonic wars, which had to be sailed right round the world even though they kept springing leaks and losing masts. The process of repairing and maintaining them had to be an integral part of the voyage, otherwise they never would have stayed afloat. The difference with the NHS is that the challenges facing it keep changing, to the point where we're no longer just talking about repairs and maintenance, but complete redesign: like trying to turn a wooden warship into a computerised starship while it's still afloat.

For these reasons leadership and change management are really important. To some extent it comes back to organising your time properly again: if you don't give yourself enough time to find out what's going on, and think through the issues involved, you won't be able to make sensible or ethical decisions about how to move forward.

But it's also about leadership – which doesn't necessarily mean one person taking control and calling the shots. There are many different models of leadership, but luckily I can only remember a few:

- Charismatic leadership (eg. Adolf Hitler – good at getting things done, not always so good at doing the right things)
- Transactional leadership (you scratch my back and I'll scratch yours – or, you do my evening surgery and I'll do your visits – always a useful skill)

- Transformational leadership ("we're not sailing a wooden boat any more, chaps, we're piloting a starship – so we're all going to have to change and develop and learn new things")

Transformational leadership is especially relevant to the NHS. It's about creating an environment where everyone can develop and achieve their potential. You can make people change their ways by issuing commandments and threatening them with punishments if they don't comply: or you can make collective decisions about the way forward, and learn new things together, so that in the end you all feel more fulfilled in your work and that the changes you have made have brought about real improvements in the things you do. This is the idea behind transformational leadership.

Domain 2 – Person-centred care

Okay, you've already got a great big list of subjects in the curriculum, on top of which you've got six domains and three essential features. You might think that's enough lists to be going on with, but when it comes to person-centred care there's another one, devised by somebody called McWhinney (IR McWhinney, *A textbook of family medicine*, Oxford, 1997). He identified three core principles of patient-centred care – namely

1. Committing to the person rather than to a particular body of knowledge
2. Seeking to understand the context of the illness
3. Attaching importance to the subjective aspects of medicine
4. Continuity of care

Oops, that's four. McWhinney doesn't actually put continuity of care on his list, but it turns out to be just as important as the others, so he might as well have done.

Basically person-centred care is all about trying to decide, along with the patient, what is appropriate for him or her as an individual, rather than just being guided by what the latest textbooks and current protocols happen to say.

Let's imagine, for example, that a woman in her eighties with known heart problems has a raised cholesterol result. You suggest to her that she ought to be on a statin, because the guideline is to treat people with statins if they are on the CHD or diabetes registers and their cholesterol score is above 5.

But the woman says she doesn't want to be on a statin. She's on plenty of tablets already and doesn't want to be on any more. She's in her eighties, so statin treatment isn't going to significantly prolong her life-expectancy. She eats the occasional cream cake, and she'd rather continue to do so, rather than spend the remainder of her days denying herself cream cakes for the sake of making the cholesterol figures on her medical record look better.

Her preferred course of action would be not to have any more cholesterol checks, even though the protocol for people on the CHD register is to have cholesterol checks once a year. She knows that if she has her cholesterol checked it's probably going to come back high, and she knows that if it's high she'll worry about it, because that's what she's like. But she also knows that she won't want to go on a statin or stop eating cream cakes. So she'd prefer to stop having the blood tests.

What's the right answer here? There isn't one, but it's clear that if you think about the needs of the person, the context of her age and attitudes, and her subjective feelings about cream cakes and statin tablets, then you'll arrive at quite a different conclusion from the one you would have reached if you had only considered the clinical guidelines.

So where does continuity of care come in? Well, imagine that you work in a big group practice. You and the patient have a good discussion about the issues above, and having gone through the medical considerations with her you agree to stop calling her for cholesterol checks in the future. The trouble is, unless you communicate this information to the other members of your surgery team, the Admin department will keep sending her cholesterol invites, and if she comes in and sees one of the other doctors she'll end up being told that she really ought to be on a statin. Very frustrating for her as a patient.

One solution to this problem is that every time she comes back to the surgery she can always insist on seeing you, because you're the one that understands. Lots of patients tend to gravitate towards this solution, especially if they've got complicated medical histories, because they want continuity of care. But this system falls down, of course, if you happen to be off sick, or on holiday, or simply overbooked so that you can't be seen for a long time.

Continuity of care shouldn't mean personal availability at all times, because ultimately you can't guarantee that to your patients, and in any case it's bad policy to have them relying on you to such an extent. Instead it should mean good organisation, good communication, good record-keeping and a shared methodology between you and the other members of the surgery team.

Domain 3 – Problem-solving

Problem solving is a crucial part of GP training because in general practice you need to adopt a problem-based approach rather than a disease-based approach.

One of the things this means is that when a patient comes to see you with a problem, you should resist the temptation to jump to conclusions about the diagnosis. According to somebody called Marinker (Marinker M and Peckham PJ, *Clinical Futures*, London 1998), when hospital specialists see patients they try to

- Reduce uncertainty
- Explore possibility
- Marginalize error

By contrast, when GPs sees their patients they are more inclined to

- Tolerate uncertainty
- Explore probability
- Marginalize danger

Let's imagine that you're a neurologist and a man comes to see you with tingling in his fingers and toes. Because the man has been referred by the GP, and you feel that you can trust the competence and good judgement of the GP (we're drifting into the realms of fantasy here), you feel safe to assume that it really is a neurological problem and proceed on that basis.

You would anyway. You're a neurologist. That's all you ever do. If someone gets a pea stuck up his nose you think it's a neurological problem and refer him for a brain scan.

But now let's imagine you're a GP and the same man comes to see you with tingling in his fingers and toes. You're just clicking your biro to refer him to the neurologist when you notice from his previous consultations that he saw one of your colleagues a couple of weeks earlier, with pains in his feet, and your colleague started him on Ibuprofen.

Hm. Ibuprofen can sometimes cause tingling in the fingers and toes. Why not take him off the Ibuprofen, and see if that does the trick? If it doesn't, you can always refer him to the neurologist later on.

This is an example of what's meant by tolerating uncertainty and exploring probability. It's also an example of incremental investigation ("we'll try this, and if it doesn't work then we'll try that") and using time as a diagnostic tool ("give it a week, and if your symptoms clear up then it was probably just the Ibuprofen").

But when you're working in this way, you always have to remember to use your judgement and intervene urgently when necessary. If the same man came to see you complaining of tingling in his fingers and toes, and while he was talking to you he happened to mention that he'd just lost all the feeling below his waist, then taking him off the Ibuprofen and giving it a week might not be the appropriate course of action.

Decision-Making

Decision-making is a vital part of problem-solving:

- You won't be much use to your patient if you can't make a decision
- You will make lots of decisions every day as the GP
- But decision-making is complicated because you have to find a balance between evidence based medicine (EBM) and values-based medicine (VBM)

So are there any guidelines? Yes there are ethical guidelines –

- consequentialism -- the action is right if the result is good ("I gave the patient the wrong treatment but they got better anyway")
- deontology -- ("the action is important, not the consequences") ("the operation was a success, but the patient died")
- utilitarianism – the best course of action is the one which results in the most good

As you can see consequentialism and deontology don't agree with each other. Utilitarianism may seem like a convenient way out of this difficulty, but the problem with a system based on measurements and comparisons is that it may not be able to take account of things like people's feelings. How is "the most good" measured? Does it mean happiness or utility? If we only take account of things which can be calibrated, then we may be forced to the conclusion that a skinny morose patient who neither smokes nor drinks is a better outcome than a red-faced cheerful old fatty who's always boozing and smoking big cigars. It's at least possible that patients who occasionally overstep the boundaries of medical advice may sometimes enjoy themselves more than those who spend all their time worrying about their blood pressure and avoiding things which might be bad for them. (Ah… whither happiness?)

Remember Dr Hairy's first rule of decision-making: "just because you can do something, it doesn't mean you should". Just because you can drink a whole pint of beer in less than 10 seconds whilst balancing on one leg with a sausage up your nose, that doesn't mean it's either clever or advisable. Likewise, you may be able to lower someone's cholesterol or blood pressure by putting them on expensive tablets, but is it always the right thing to do?

In general practice we're not always dealing with things which are clear-cut and quantifiable; just as often, or perhaps more often, we're dealing with things which are complicated and uncertain. So how do we make good decision under these circumstances?

Dr John Gillies, author of the excitingly-titled *RCGP Occasional Paper 86*, suggests an approach called "practical reasoning", which combines the following:

- priority of the particular (concentrate on the particular situation rather than rules)
- situational appreciation
- practice of perception
- subjectivity of interpretation
- EBM (our underlying science)
- intuition

You may wonder what subjectivity and intuition are doing on the list. Surely decision-making should be a rational process? Well, perhaps it should be, but functional MRIs show activity in the amygdala during decision-making, and there is evidence from people with damage to their limbic systems that decisions cannot be made without emotional input. In the real world – and particularly in the crucible of the consultation – decision-making is not a matter of pure logic.

It doesn't follow from this that you should just go with the flow (like a bobbing cork), let it all hang out (like a man with his flies undone) and do whatever your feelings dictate. You have to engage emotionally in order to make good decisions, but having done so you have to be appreciative, perceptive, self-aware, knowledgeable and practical in order to translate this engagement into something which is really useful to your patients. Being aware of your own values is also important, since these are bound to influence the decisions you make. Above all, you will need practice to develop your awareness and understanding of your patients, and hence your decision-making skills.

Decision-making II

Decision making doesn't just take place in the consultation: as a GP you will have to make lots of decisions about how to run a practice, how to respond to changes in policy, how to alter practice when new guidelines are issued, where to go for the staff Christmas party, and loads of other stuff as well. Probably some of these decisions will be made informally over a cup of tea, but others will have to be made at formal practice meetings. There will be times when people have to be told that they're doing things wrong, and other times when you have to face the fact that you've been doing something wrong yourself. The process of getting things done when working with other people can sometimes be an uncomfortable one, but on the other hand you'll never make an omelette if you're frightened of cracking eggs.

Here's an example from my own practice. I'm a Course Organiser, and on Wednesday afternoons I'm usually at the local Postgraduate Centre training Registrars.

Since we're a single-handed practice, this means we have to find cover for the afternoon surgery. In the past we've made do with a patchwork of locums, but it's never been satisfactory. They don't know our protocols, there's no continuity of care, and of course they're very expensive.

Last year, however, we were lucky enough to get a Retainer GP called Sally who came and worked for us all day on Wednesday, covering the afternoon on her own. She has four school-age children, but her mother agreed to collect them from school and take them home for tea every Wednesday.

This summer, unfortunately, the mother suddenly announced that she was too old to do this any more: so Sally had to give the Wednesday afternoons up.

We thought we were going to have to go back to employing locums, until I came up with the idea of swopping half-days. At the moment our half-day is on Thursday, and on Thursday afternoons another local surgery, run by Dr Justin, covers us for emergencies. In return we cover for Dr Justin on Wednesday afternoons, when he takes his own half day. But if we were to swop over – so that we closed on Wednesday afternoons and he closed on Thursdays – we would no longer need cover for my training sessions.

Rather to my surprise Dr Justin agreed to this idea straight away. Splendid, I thought, all my problems are over (apart from that unfortunate matter with the local Masons) – but then I started to realise that my decision would have other knock-on effects, which I hadn't anticipated.

For one thing we rent out a treatment-room, and on Wednesday afternoons it's used by a private physiotherapist. Now she'll either have to change to Thursdays or find herself another room somewhere else.

For another thing, on Wednesday afternoons our Reception duties are looked after by Jill, one of our best members of staff, and she can't swop to a Thursday because she's got another job at another surgery. She wasn't very happy about having her hours reduced, and it was only after talking things over with her at great length, assuring her that I didn't want to upset her and trying all sort of ideas to work out how she could make up the hours, that we came up with the idea of her staying late on the Wednesday, in the quiet, to do some scanning and typing.

This seems to exemplify the ways things happen in general practice. Many of the problems you have to deal with come about because of factors outside your control, like Sally's mother. You have to try and solve these problems by thinking flexibly and creatively about your working arrangements; but whatever decision you take, it's bound to create secondary consequences that you haven't foreseen. (Ah, ripples within ripples…)
Eventually, by talking things over in depth with the people concerned, you'll find your way to a workable solution – just in time for the next problem to plop into your lap.

Domain 4 – Comprehensive approach

Interpretation

Most GPs start their consultations with a general phrase such as "What can I do for you?", which is a way of inviting patients to explain why they've come to see you and what they expect you to do for them.

It's very unusual, in response, for the patient to say something like "Well, doctor, I've got a number of different problems. There's what looks like a solar keratosis on my forehead, I've got a cyst on my epididymis, I'm experiencing unpleasant clicking sensations in my left knee and I'm retaining fluid around my ankles, although I think this is probably secondary to my heart failure."

Patients often have multiple complaints, but they don't always differentiate between them. They're just as likely to say "When I was at home yesterday I heard a ringing noise, then everything went black and the muscles of my left leg went very tight. Now every time I look at a map of the British Isles all the roads south of Watford look a bit yellowish. My next door neighbour says her husband had that, and when he had a scan it turned out he'd swallowed a thirteen-amp plug."

The challenge for the GP is often to extract relevant information from the patient and interpret it, or decide how to investigate it further. If there seem to be multiple problems, then they may need to be prioritised in consultation with the patient ("I think we should deal with the hatchet in your skull first, Mrs Smith, and worry about your protruding ears a bit later").

Ongoing care

Once you've established what's the matter with the patient, you've still got to carry on taking care of him or her. After all, there are a lot of problems which can't be fixed even when they've been identified and thoroughly investigated and the appropriate referrals have been made. Deciding that a patient has Alzheimer's is only the beginning of a care process.

A lot of general practice isn't about making a diagnosis and then curing people: it's about the same patients coming back to see you over and over again, with pretty much the same problems, about which only a limited amount can be done. Under these circumstances, patient care is about monitoring and controlling your patients' conditions, rather than curing them. Things like diabetic checks are a classic example of this kind of care.

On the subject of diabetic checks, it's now recommended that diabetic patients should have depression assessments, because a chronic disease such as diabetes can often lead (understandably) to depression. This is a fairly good example of the comprehensive approach: thinking about your patients' illnesses in the broader context of their lives, instead of dealing with each condition in isolation. This will also help you to relate to your patients as individuals, instead of thinking about them as medical conditions on legs.

But a comprehensive approach to patient care also involves promoting the health and general well-being of your patients, whether they're ill or not. If a man comes to see you about an ingrowing toenail, it may be worth your while measuring his height, weight and blood pressure, recording his alcohol consumption, asking about his smoking status and giving him the appropriate lifestyle advice while you've got him there, because apart from improving his long-term quality of life you might just save the NHS a lot of time and expense further down the road. Then after he's gone you can light up a self-congratulatory panatella, help yourself to a slice of chocolate cake and pour yourself a large whisky, on the strength of your good deed.

The GP contract has an important part to play in ongoing care, because it identifies a number of really common chronic diseases – CHD, diabetes, epilepsy and so on – and provides a structured framework for their ongoing care and monitoring. Here are a few examples:

- Demented patients should have general health checks, because research shows that they suffer from common physical problems as much as anybody else, but these tend to go unnoticed because of the dementia
- Diagnoses of asthma should always be confirmed by spirometry or reversibility
- Diagnoses of depression should always be confirmed (and the severity of the depression established) by the use of an approved questionnaire
- Patients with a new diagnosis of cancer should always be offered the opportunity to talk over the diagnosis, care pathway and likely outcomes within a few weeks of the diagnosis being made

These are all good ideas. But the problem with the GP contract is that, because it offers points to surgeries if they tick all the right boxes, and these points get translated into money at the end of the year, it encourages you to

- Think about patient care as a box-ticking exercise – which means that you think about the things which get you points, and maybe overlook other things which don't, even if they would be good for your patients.
- Regard patients who lose you points (for example diabetics with high HbA1c, or CHD patients who don't want to use a statin) as a terrible nuisance, and get very grumpy every time you see them.
- Suddenly remember, after your COPD patient has gone, that you forgot to observe her inhaler technique – then tick the "Inhaler technique observed" box on the template, fraudulently, because you can't bear the thought of losing the points.
- Draw up a list of asthma patients who haven't come for their flu vaccines, lurk around outside their houses late at night with a high-velocity rifle, and pick them off when they emerge.

There may, in other words, sometimes be a conflict between highly-structured care and comprehensive care. But that's no excuse for shooting your patients.

Co-Ordination of Care

When patients have lots of complicated problems, they often become involved with lots of different agencies, such as

- The District Nurses
- The local Respiratory Team, if patients have COPD
- Oncology, if patients have cancer
- The Diabetic Specialist Nurse, if patients have diabetes
- The local Hospice, if the patient is terminally ill

These are just examples, of course. It's not at all beyond the realms of possibility that a single patient could be involved with all of these different teams. What this means is that even if the GP isn't all that heavily involved with the patient on a day-to-day basis – which may well be the case if the patient has limited mobility and therefore doesn't get into the surgery very much – there may still be an important role to play, with respect to co-ordinating care. For example:

1. You ask the district nurse to go and see the patient to carry out a blood test
2. When the blood test comes back it's raised, so you ask the district nurse to repeat it in two weeks
3. You also send a copy of the blood test to the Oncology Department, which is now the patient's main point of care
4. Two weeks later you get a fax from the Out of Hours service to say that the patient was taken into hospital overnight
5. Later that day, the district nurse rings up to say that she can't do the patient's repeat blood test because the patient won't open the door. Is the patient all right? Has the patient died?

Number 5 is your fault, because you knew that the patient had gone into hospital but you forgot to pass on the information to the district nurse. Of course you were terribly busy at the time when you saw the fax and it's very difficult to remember every little thing – but this is an example of how complicated it can get co-ordinating people's care when there are lots of different agencies involved.

Domain 5 – Community orientation

The work of family doctors is largely determined by the makeup of the community in which they practice. Let's take an extreme example: one of our Registrars once brought her father to the surgery to meet us. He was the sole GP of a small town near Hyderabad in India, population 54,000. He told us that he usually saw about 150 patients per day, and most of these consultations consisted of giving vaccinations against water-borne disease. He never saw any patients with alcohol or drug dependency.

Perhaps it's not surprising that the work of a GP in rural India should be very different to the work of a GP in the UK; but even in the UK there are big differences between one community and another – between a urban communities and rural ones, for example, or between a posh urban community like Mayfair and a much less posh one like Maidstone.

In all communities healthcare resources are limited, which means that doctors inevitably become involved in rationing decisions. This means that doctors have an ethical and moral duty to influence health policy in their communities, and that when they are dealing with individual patients they need to balance the needs of the individual with the needs and financial constraints which apply to the community as a whole.

Let's imagine that you were working in a community of pirates:

- Every Saturday night the pirates would have too much grog and a huge fight would break out in the town centre, with the result that large numbers of them would end up in A&E with cutlass wounds.
- The local Ophthalmology Dept would spend most of its budget on black eyepatches.
- The local Appliances Dept would be overwhelmed with requests for wooden legs and hooks.
- You most common referrals would be to Dermatology, for skin-diseases caught from parrots, and to Orthopaedics, for shivered timbers.

You'd spend a lot of your time explaining to pirates that the waiting-time for a new wooden leg was bound to be long, because there was such demand for them. And when the local Practice Based Commissioning group began to investigate how to reduce rates of admission to hospital, they would want you to try to persuade your patients to moderate their grog intake on a Saturday night. But they never would reduce it, blast your eyes. Be damned if they wouldn't rather roast in hell than go without their grog. Any more of that lily-livered talk, and by thunder they'd string you up from the nearest yard-arm.

As you can tell from this example, the makeup of the community where you work is bound to exert a powerful influence over

- The kinds of illnesses you see most frequently,
- The kind of things that the health budget for your area ends up being spent on, and therefore
- The kind of preventive medicine which might make the biggest difference in your area

But if only those pirates would tell you where the gold was buried, all your health budget problems would be over.

Domain 6 – holistic approach

Holism is a very important concept in the curriculum, but what does it actually mean?

- Pretending to listen to a patient but actually thinking about your hols?
- Peeping at your patients through a hole in the wall?
- Becoming prejudiced about your patients if you don't like the look of their holes?

Nope. It actually means --

- "Caring for the whole person in the context of the person's values, their family beliefs, their family system, and their culture in the larger community, and considering a range of therapies based on the evidence of their benefits and cost"
- Taking into account "inner experiences" that are subjective, mystical (and, for some, religious), which may affect your patients' health beliefs
- Patient-centred care, which can be helped by longer consultations (giving the patients time to talk things out thoroughly) and greater continuity of care (encouraging the patients to feel more familiar with the care they are being offered, giving yourself more of a chance to get to know the patients as individuals)

Holism involves an acceptance that there's more to health and wellbeing than the things medicine can measure.

The novelist Thomas Hardy tells an interesting story about an old man who is confined to his bed and can see, through his bedroom window, a big tall tree which grows just opposite. The tree is quite old and creaks and sways about alarmingly every time the wind blows, and the old man spends all his time worrying that the tree is going to fall on his house. He never talks about anything else. A new doctor comes to the village, and, deciding that the old man's health is being ruined because of his constant worrying about the tree, he orders the tree to be cut down, despite the quavering protests of the old man himself. As soon as the tree is down, however, the old man goes into a terminal decline. Without the tree, his life no longer has a focus. He dies shortly afterwards.

What does this story illustrate?

- Thomas Hardy likes a tale with an unexpected twist to it.
- When you get old, you should try to avoid being confined to bed in a room with a big tree outside the window which creaks and sways about alarmingly every time the wind blows.
- Good doctors understand their own limitations, and tailor their treatment to fit the patient – "patient-centred care" – instead of forcing their ideas onto people.
- As a doctor, you need to remember that people's inner lives are complex, and sometimes what may look like an obvious solution to you may not work for the patient because of complex psychological factors.

If you see a grossly obese woman with bad knees, the obvious thing to do is tell her to lose weight, because the weight is probably causing the knee problems. Then when she comes back to see you two months later, heavier than ever and still complaining of knee problems, you think she must be stupid. But there may be all sorts of complex background reasons why she finds it difficult to diet:

- She's lonely and depressed and therefore given to "comfort eating".
- Everyone in her family is overweight, and if she goes round to see her mum or her sister for tea they're bound to be eating cakes and chocolates, and she finds it very difficult to be the odd one out.
- Her husband discourages her from dieting, because he's overweight himself but scornful of medical advice.
- She works as chief taster in a fondant fancy factory.

Holistic care means getting to grips with some of these background factors as well as the obvious foreground issue of your patient's obesity.

The Three Essential Features

Contextual Aspects

General practice doesn't take place in a vacuum: it overlaps with lots of other things, like one of those Venn diagrams with lots of overlapping circles. Here are a few examples of the other things with which general practice overlaps:

- The local community – if it's a well-to-do community, for example, then you're likely to make a lot more private referrals
- The local hospitals – perhaps one hospital has a particularly good gynaecology department, another one is particularly good at cardiology, and there are short waiting-times for orthopaedics at a third
- The Primary Care Trust – what kind of enhanced services are they running? Does your practice participate? Do you even know what an enhanced service is?
- NICE – what's the latest advice on how to deal with hypertension? What kind of thing should you prescribe for smoking cessation, and under what conditions?
- Financial regulations – Who is your accountant? Is it wise to keep all your receipts in a shoebox under your bed? How come your annual salary looks impressive, but your bank account is always in the red?
- Health and safety regulations – If you have a trapdoor rigged up in your consulting room, which drops open when you pull a lever, hurling particularly irritating patients into a pit of ravenous crocodiles, does this contravene any regulations?

- The legal system – Was it okay to fraudulently claim for 2,000 flu vaccinations last year, when actually you only gave 200? They can't touch you for it, can they? What's that blue flashing light outside the surgery?

Remember, these are only examples. No GPs in real life would ever let their bank accounts get into the red.

Attitudinal Aspects

This doesn't mean, unfortunately, that the new curriculum would like to encourage you to be a GP with attitude, hangin' with the kids, takin' it back to the streets, etc.

What it means is that your own attitudes will play a big part in your life as a GP, and it's therefore very important that you should be self-aware about your own capabilities and values.

You need to be able to identify those aspects of clinical practice which involve ethics and values, and sort out how your own personal ethics and values will influence your way of dealing with them. For example, if you're a strict Roman Catholic or Muslim, how will you feel about dealing with underage sex, requests for the morning-after pill, or abortions? How do you feel about dealing with drug-addicts?

You also need to be aware of the interaction between work and private life and attempt to find a good balance between them. You don't want to become so obsessed with your work as a GP that you've got no outside interests whatsoever, because sooner or later the job is going to get you down, and you'll need some way of relaxing or letting off steam.

On the other hand you don't want to become so involved in all-night parties and beer-drinking competitions that you only get to bed at five o'clock in the morning and then find it difficult to concentrate during morning surgery, because your head is throbbing, the room is going round and there's a green haze in front of your eyes. A good balance between work and play is essential.

Another area where it's important to retain a balance is finance. GPs are pretty well paid compared to the population at large, but that doesn't stop them from getting into financial difficulties. And it's comparatively easy to stay on the moral high ground when you've got a healthy bank account, but after your third divorce, when the kids are at university constantly demanding more money for books and clothes, and the mortgage on your new house is more than you can really afford, fiddling the QOF points suddenly starts to seem much less morally repugnant. Or so I gather. I mean, it might do. You can imagine what that might be like.

Scientific aspects

After you've been a GP for some years you'll get so used to giving patients placebo prescriptions, for the sake of a quiet life, that you might start to forget that medicine is actually meant to be based on science.

The other thing which will happen is that you'll see so many useful pamphlets from NICE about how to handle hypertension, depression, respiration, abrasion, subtraction, multiplication and King's Cross station, that you'll stop reading them. In fact if you're not careful you'll stop reading any new guidelines or any new research whatsoever.

Medicine doesn't stand still: research goes on all the time, and new evidence is always being brought to light, which often results in a redefinition of "best practice". But it's also true that you can't possibly expect yourself to keep abreast of every single new development, because you'd never have time to do any work, get to bed or even enjoy that all-important midmorning coffee and biscuit.

Be that as it may, however, you can't afford just to turn your back on the whole thing. It's important for your personal development (not to mention your reappraisal) that you should remain interested in what's going on scientifically. You need to be familiar, and remain familiar, with things like the methods and concepts of scientific research and the fundamentals of statistics; you have to be able to read and assess medical literature critically; and you must be able to maintain continued learning throughout your career as a doctor.

This is one area where communication with your colleagues in general practice can be invaluable, as can learning through teaching. It may be impossible for an individual to keep abreast of all the latest developments, but a group of you can do a fairly good job of it, either by giving presentations to each other or simply by talking things over amongst yourselves.

You can even get ahead of the game. Try boning up on some really obscure subject like bisphosphonates. Then when you get to work, introduce the subject into conversation at the earliest opportunity. "I say, Smithers, have you seen the latest advice about bisphosphonates? Jolly exciting, isn't it?"

Notice the look of panic which comes into Smithers' eyes. No, he hasn't seen the latest advice about bisphosphonates. To tell the truth, he isn't even completely sure what a bisphosphonate is. Isn't it some kind of luminous undersea creature, with its shell divided into two halves? "Er, yes," he mutters, "we must have a chat about that some time... I'm a bit busy just at the moment, though." Then he hastens to his consulting-room.

From now on, everybody will assume that you're dead keen on medical research, and they'll always defer to your opinion when the question of bisphosphonates comes up. (Should you serve them with a green salad or a seafood sauce? Dr Hairy's bound to know.) Pretty soon you'll be regarded as the bisphosphonate expert for the whole county. The Nobel Prize for Bisphosphonates could one day be yours. So you see, medical research really is worthwhile, after all.

Core and Extended Statements (The Big List of Subjects)

The purpose of these Statements is to detail how the Six Domains and Three Essential Features can be applied in a variety of different contexts and what learning resources are available. In other words, this is where you have to get down to the nitty-gritty and demonstrate that you don't just understand the theory and philosophy of General Practice: you can apply this theory and philosophy to the details of real-life patient care and management.

Each Statement has a number of learning outcomes which are supposed to show that you understand how to apply the Six Domains and the Three Essential Features to the area covered by that Statement. Altogether there are over a thousand of these learning outcomes (aagh!), but you don't have to do them all (phew!). You're supposed to do enough to demonstrate that you can apply the Six Domains and the Three Essential Features in a representative variety of contexts. Normally, about nine hundred and ninety-nine learning outcomes will be sufficient. Only joking.

Here is the complete list of Core and Extended Statements. There are 32 of them, organised into 15 groups:

1. The "Core Statement" – Being a general practitioner
2. The general practice consultation
3. Personal and professional responsibilities
    3.1. Clinical governance
    3.2. Patient safety
    3.3. Ethics and values based medicine

- 3.4. Promoting equality and valuing diversity
- 3.5. Evidence based health care
- 3.6. Research and academic activity
- 3.7. Teaching, mentoring and clinical supervision
4. Management in primary care
    - 4.1. Management in primary care
    - 4.2. Information management and technology
5. Healthy people: promoting health and preventing disease
6. Genetics in primary care
7. Care of acutely ill people
8. Care of children and young people
9. Care of older adults
10. Gender-specific health issues
    - 10.1. Women's health
    - 10.2. Men's health
11. Sexual health
12. Care of people with cancer & palliative care
13. Care of people with mental health problems
14. Care of people with learning disabilities
15. Clinical management
    - 15.1. Cardiovascular problems
    - 15.2. Digestive problems
    - 15.3. Drug using adults
    - 15.4. ENT and facial problems
    - 15.5. Eye problems
    - 15.6. Metabolic problems
    - 15.7. Neurological problems
    - 15.8. Respiratory problems

15.9. Rheumatology and conditions of the musculoskeletal system (including trauma)

15.10. Skin problems

I haven't attempted to comment on all of the Extended Statements in detail, otherwise you'd end up with another document 600 pages long, like the New Curriculum itself, and you'd be right back where you started. Below are a few general remarks about statements 6-15, which deal with Medicine, followed by some specific (and brief) notes about individual sections.

Medicine (statements 6-15)

The following is a brief summary of some of the key areas in these sections:

Neurology
- Epilepsy
- Emergencies, meningitis, subarachnoid haemorrhage, loss of consciousness, stroke
- Appropriate referrals/investigations/treatment
- Chronic disease management – Parkinsons, MS, dementia
- DVLA
- Attend neurology clinics
- Headaches (feel one coming on…?)

Respiratory
- Smoking cessation
- CDM – asthma, COPD and emergency management
- Antibiotics
- Symptoms, conditions, investigations (PEFR, spirometry), emergency (anaphylaxis), prevention

Rheumatology
- Back pain/neck/shoulder/knee (etc.)
- Soft tissue disorders
- Osteoarthritis
- Osteoporosis
- Inflammatory arthropathies
- Polymyalgia rheumatica, temporal arteritis

Trauma
- Investigation, treatment, allied health professional roles, CDM, prevention, exercise and accident prevention
- Shared care
- Majority is dealt with in general practice

Acutely Ill
- Teamwork/leadership
- Prioritising
- OOH (Out of Hours)
- Appropriate referrals
- Autonomy/continuity
- Review and safety-netting
- CDM/co-morbidity/prevention
- Carers/support/bereavement
- Mental health act/compulsory admission
- EBM/protocols/significant event analysis
- Resuscitation certificate yearly

Genetics
- 1:10 patients has disorder with genetic component
- Understanding inheritance patterns
- Draw family tree
- Good communication
- Antenatal screening

Cardiovascular problems
- Management of risk factors and emergencies

       Appropriate management and referral
       NICE guidelines
       DVLA

Digestive problems
       Dyspepsia & GORD management/EBM
       Colorectal cancer
       Indications for appropriate and urgent referrals
       Acute abdomen/GI bleeds
       IBS (Irritable Bowel Syndrome)

Metabolic problems
       Obesity and diabetes mellitus; NSF & GMS2
       Thyroid disease
       Adrenal emergencies/diabetic emergencies
       Gout
       Coordination of care

How does this differ from a hospital orientated curriculum or training, you may ask – surely there is a huge crossover? After all medicine is medicine, there are only so many diseases and conditions, and you can't divide these up neatly into the ones which are dealt with in hospital and the ones which are dealt with in general practice. So why bother with a special curriculum for GP Registrars?

The key difference is your view-point or where you stand, and this is why it is important to relate learning and teaching of the Registrars' Curriculum to the six domains and essential features. For example, if you're a rheumatologist and a man comes to see you with a pain in his foot, you investigate him for arthritis. If you're a vascular surgeon, you investigate his poor circulation. If you're a diabetologist you think the pain in his foot is probably secondary to his diabetes and treat that. If you're an orthopaedic surgeon you chop off his foot and fit a false one.

If you're a GP, on the other hand, you have to consider all these different aspects and possibilities, and you also have to listen to the patient's own theory that he's got a brain tumour, because that's what his uncle had and he started off with a pain in his foot. Then you suddenly realise that the patient's shoes are too tight.

In other words, the New Curriculum may cover a lot of the same ground as hospital training, but it covers it in a different way, by relating it to the Six Domains and three Essential Features – the core values of general practice.

The general practice consultation (statement 2)

The consultation is at the heart of general practice. We need to be able to communicate properly with patients if we are going to be able to do all the six domains and three essential features stuff. If you don't understand your patient's problem fully and in context you can't practice holistically or be patient centred.

So ... good communication is needed for -

- Developing a partnership between doctor and patient
- Trust and openness
- Shared decisions and understanding.

These things don't just apply to communication with patients but are similar to what goes on in teaching and learning and are also useful when communicating with colleagues and team members.

Context is very important and forms quite a big list of learning outcomes in this section of the curriculum (go on, have a look).

Here is a shortened version of this list -

- Recognize diversity
- Respond flexibly
- Share understanding
- Think about relatives and friends (the patient's not yours – well, not during the consultation anyway)

- Know about consultation models and think physical, social and psychological when considering the diagnosis

Recognizing diversity is important because we are dealing with people and not just diseases. Everyone experiences their ill health in different ways and the effects of ill health have different meanings for each individual. Or, "people step in the same rivers and different waters flow over them ".

Knowing about consultation models is not just a theoretical thing, it is useful practically. Models are a bit like recipes or route maps. They help you to know where you're going in a consultation and help you find your way if you get lost. But you also have to learn how to think for yourself, and above all how to pay proper attention. Try not to think about your supper when patients are telling you about their piles. You'll only end up referring them for pies, or having a big bunch of grapes for dessert. Here's a story -

> Tich Nhat Hanh and his friend were sitting under a tree having a picnic. The friend was eating a satsuma and putting another segment in his mouth before he had finished the first while talking excitedly about his plans. Tich Nhat Hanh pointed this out to him and suggested he should concentrate on one segment at a time.

The moral of this story is that you should concentrate on the moment. And never let your patients eat satsumas during their consultations.

Ethics and Values Based Medicine (VBM) (statement 3.3)

This is about "a process of working with complex and conflicting values in healthcare in an attempt to produce the best decisions". So this chapter is partly about ethical principles, values and decision making. It is also about holism: big picture stuff. It's about the balance between –

    evidence-based medicine   values/ethical principles
            doctor   patient

It's also about the partnership between doctor and patient.

This balance is the process by which generalists (or holists) make decisions -- which is good because this is how the brain works to make decisions (in A Hairyman's opinion).

Values based medicine is made up of four things (less than a handful) –

1. awareness
2. knowledge
3. reasoning
4. communication

Values based medicine deals with complexity of conflict, eg. –

| values based medicine | *versus* | clinical governance |
| budgetary controls | *versus* | quality of care |

So back to the four things that make up values based medicine –

Awareness

- Of diversity of values (patients, doctors, society)
- of gaps (between patients, doctors and professional disciplines)
- of embedded values eg. in policy and the guidelines
- of personal values (and how these affect our decision making)
- of uncertainty arising from gaps (see above)

Reasoning

- understanding ethical codes/guidelines/law
- understanding "decision theory"
- principles of ethical reasoning (utilitarianism, consequentialism)
- practical use of methods to reach a decision
- openness (two different values, respect)

Knowledge

- sources eg. patient narrative
- principles into practice (policy)
- continued professional development (considering different values, different ways of knowing, different ways of seeing)

Communication

- communication skills (patients and team work)
- reflection, partnership, negotiation

Self awareness is very important in balancing values, ethics and evidence, and reflection is very important in developing self-awareness.

Conclusion

- be open, imaginative, flexible and understanding
- understand yourself
- know about decision theory (see above, under Domain 3 - Problem Solving)

Evidence-based health care (statement 3.5)

This is about providing information to patients to help them make the right decisions. You need to be able to –

- turn uncertainty into a question
- retrieve evidence
- appraise evidence
- evaluate the outcome (Sicily statement)

There is a balance between –

evidence-based medicine (EBM)	values based medicine (VBM)
randomised controlled trials	autonomy and equity
science	culture, philosophy, spiritual

This balance is important as we must recognize the limitations of EBM and accept that there are more unknowns than certainties.

Most research is done in secondary and tertiary care and may not be directly applicable to patients in primary care, especially when exclusion criteria and comorbidity are considered.

EBM fuels change and is, therefore, linked to leadership and policy making. However, a lot of research is funded by drug companies which in turn affects resource allocation and may result in disadvantaged groups of patients.

Conclusion

- It's a balance

Information Management & Technology (IM&T) (statement 4.1)

In the old days IM&T used to be called just plain IT, but this was too short and didn't sound important enough, so about five years ago the NHS decided to call it IM&T instead.

Computer training

It may be very difficult to find the time for any formal computer training, especially as you've got such a lot of training to think about already. On the other hand, there's no prospect of general practice, or indeed anything else, ever reverting to a non-computerised method of work, so computer training might be one of the most useful pieces of learning you could ever undertake. The most widely-recognised basic qualification in computers is the ECDL, or European Computer Driving Licence. It isn't difficult, but it's quite wide-ranging: in order to pass it you need a broad knowledge of files and folders, memory measurements, word-processing, computerised spreadsheets, relational databases, computerised presentations and e-mail. Starting from scratch, it's estimated to require about 80 hours of study.

Read Codes

Read codes were introduced as a means of standardising patient records, and they are now used by all primary care clinical systems. They are combinations of letters and numbers: Ischaemic Heart Disease is G3, for example. There are thousands of Read Codes, divided into chapters, including separate chapters for

- History
- Examination
- Procedures
- Diagnoses
- Medication

One problem with Read codes is that although they're used throughout General Practice they are not used in hospitals, which have their own coding-systems. As the electronic exchange of information between surgeries and hospitals becomes more and more prevalent, therefore, it will become increasingly important for both sides to use the same language, and the Read system is likely to be replaced by something else.

Communication/interaction

How do referrals get from the GP to the hospital? On paper, or via Choose & Book? How do results get from the hospital to the GPs, and how do they get from GPs to patients? Can patients inspect their own results online, book themselves appointments, or reorder their repeat prescriptions? All of these things are changing rapidly as IT within the NHS is developed.

Confidentiality

In a computerised world, confidentiality is no longer just about remembering not to discuss Mrs Brown's STI in front of her neighbour Mrs White. It's also about remembering not to draft a referral letter on your laptop, then give the laptop to your nephew without wiping the memory. And it's also about whether patients are happy to have their medical records shared electronically with the local hospitals, social services and other agencies.

Sources of information

In the old days, if you wanted to know how to recognise and treat Still's disease, you had to look in a book. Nowadays you look on Google. But how trustworthy is the Web as a source of information, and how do you know that medication recommended in the USA is going to be approved by NICE in the UK? You need to know where to find good information and approved guidelines. You also need to find this information in a way which augments the consultation and your rapport with the patient, rather than disrupting them. Try to avoid getting so involved in a huge web-search for "chest pain" that you fail to notice your patient clutching his sternum, emitting a strangled gasp and sliding to the floor in a senseless heap.

Healthy people: promoting health and preventing disease (statement 5)

Health promotion is about

- Patient self-care partnership and choice
- inequalities in health (Black report)
- patient's values and aspirations and beliefs
- wider health agenda - health of community

GPs are in a good position to promote health because of their regular patient contacts and knowledge of families.

You need to know about –

- principles of screening
- teamwork
- concepts of health, function and quality of life
- epidemiology
- principles of prevention, immunisations, and behavioural change.

Health promotion can't be imposed on patients against their will: it should arise from an educational process, a proper discussion of the issues involved, and a shared sense of empowerment and commitment.

Having said this, it must be recognized that the patient's agenda and health promotions agenda may conflict and cause tension between doctor and patient. You may well think that patients would benefit from losing weight, washing their feet, giving up smoking and changing their underpants. But if you pick the wrong time, or say it in the wrong way you may open the way for a dysfunctional consultation. Your patients may tell you that they only eat fruit vegetables and nuts and then outline their diet of the last week in detail to you. They then explain that the only reason they are overweight is because they work in a fish and chip shop and must be absorbing the vaporised fat in the deep fat fryers through their skin. They will probably tell you they are allergic to all soaps and couldn't contemplate giving up smoking at the moment as their hamster has just choked to death on a toffee, and how could you be so heartless to suggest it at a time like this? Anyway, their father smoked 300 cigarettes a day from the age of three, lived until he was 115 and played football for Scunthorpe until he was 96. Finally, the underpants were an ancient family heirloom and bestowed the wearer with great powers of patience, tolerance and kindness so luckily for you, your patient is able to forgive you your insulting suggestions.

One final word of warning; if we are going to advise someone to start medication for hypertension, warfarin rather than aspirin for atrial fibrillation , a statin for raised cholesterol or have penicillin after a splenectomy then it is our responsibility to know what the evidence is and convey it to the patient so they can make an informed decision.

Conclusions:

- Respect patients' values and aspirations and beliefs.
- Don't bully patients into a certain line of treatment; know the evidence and convey it clearly

Care of children and young people (statement 8)

You need to know about –

- child abuse and child protection
- confidentiality, communication and consent
- normal growth and development
- recognizing the seriously ill child
- access for teenagers
- immunisations
- care pathways and multidisciplinary working
- children with disabilities, mental health problems and those who have clinical or social problems that makes them vulnerable.

Gender-specific health issues (statement 10)

Men –

- take more risks with their health than women
- attend GPs less than women, because they're braver
- don't live as long as women - five years shorter life expectancy.

It's not fair, is it? Men do all the work, take all the risks, and they end up with shorter lives. In order to even out this imbalance, in a properly-organised society, men's health problems should obviously be taken more seriously than women's. For some reason, however, this is frowned upon in medical circles.

Men's health issues include –

- higher risk of heart disease
- PSA testing
- erectile dysfunction
- higher risk of suicide
- effects of social class and ethnicity on health
- rupturing themselves when they jump up excitedly when someone scores a goal in a football match

Women do have health issues of their own, however, including –

- domestic violence
- contraception, conception and high teenage birth rates
- cancer screening and referral guidelines

- legislation (with regard to termination of pregnancy and contraception for minors)
- having babies
- gynaecological problems
- swooning away from soppiness when they see fluffy animals

Women may well prefer to see a female doctor and it is important to consider their choice in this matter.

Women are often carers for other family members, and the impact of female illness on the family needs to be considered. If the woman of the house is laid low with a virus, the kids may well set the place on fire when they try to use the toaster, and the husband or partner, unable to find his own underpants, will be forced to put on women's underwear. This habit, once acquired, can be very difficult to snap yourself out of. Or so I've heard.

Sexual health (statement 11)

This is –

- government priority
- about teamwork and holism
- about contraception
- about avoiding infections
- about sexual dysfunction

Services need to be accessible, particularly to teenagers. You may need to be flexible about seeing teenagers (for example during their lunch break from school).

Uncertainty –

- national screening programme for chlamydia?
- Vaccination for HPV to prevent cervical cancer

This section illustrates the six domains and three essential features quite clearly - you really should read it.

Care of people with cancer and palliative care (statement 12)

This is an important section because cancer is common and death is inescapable. You will be involved in the care of patients from diagnosis, through treatment and palliation and finally death. But your role doesn't stop there. You need to know about death certificates and cremation forms and about supporting relatives in their grief. Continuity (in its biggest sense) and teamwork work are important and you will be your patient's advocate whoever else is providing their care.

You need to know about –

- cancer screening and prevention
- referral guidelines and early diagnosis
- palliative care principles and emergencies
- benefits and services available
- ethics, choice, advanced directives.

Care of people with mental health problems (statement 13)

You should be able to –

- Recognise depression, and distinguish it from emotional distress
- Assess suicide risk
- Recognise physical comorbidity in people with severe mental health problems
- Also recognise that people with chronic diseases may become depressed as a result of their condition (relevant to QOF, the Quality and Outcomes Framework, and CDM, Chronic Disease Management)
- Recognise and manage somatisation

Other points:

- One third of patients who see a GP have a mental health component to their illness
- Communication skills and patient centred practice are very important in the care of people with mental health problems
- This section includes psychoses, depression and anxiety, adjustment reactions and others (see Appendix 3 to this section in the Curriculum)
- NSF and NICE emphasise the need for non-pharmaceutical interventions
- You need to know about the Mental Health Act
- Don't (necessarily) be dualist
- Oh, and be holistic (see Appendix 8 to this section in the Curriculum)

Care of people with learning disabilities (statement 14)

- People with learning disabilities may have associated co-morbidities (physical, mental or behavioural)
- Mental capacity and consent need to be considered (for example before undertaking any intrusive examinations)
- On the other hand the first Key Principle of the Mental Capacity Act (2007) is that "Every adult has the right to make his or her own decisions and must be assumed to have capacity to do so unless it is proved otherwise." So you can't avoid examining people just because they have learning disabilities, or avoid calling them for smears or asthma checks or whatever. They are entitled to the same standard of health care as everyone else.
- Access to GP services and inequality are important issues in the care of people with learning disabilities.
- There may be communication difficulties; but on the other hand, communicating through a carer may affect the Doctor/patient relationship.
- You need to consider the affect of caring for the patient on the health and mental wellbeing of the carer(s).

Drug using adults (statement 15.3)

The drug use covered by this section includes alcoholism.

- There's a lot of it about
- We have a duty of care
- Treatment is available (eventually and at varying quality)
- Behavioural change is possible
- Know your local referral arrangements
- Can you spot a heavy drinker? (Try Googling "Alcoholism CAGE test")
- Learn to think about drugs and alcoholism as diagnostic possibilities during your routine work
- What is "brief intervention"? Here's a clue – it's nothing to do with persuading your patients to change their underpants. Try Googling it again.

You really need to know about this section, because you will come up against alcoholism and substance abuse repeatedly in your time as a GP.

It's important to understand your own attitudes, because they will affect how you deal with the patients. Is this an illness or a problem of society? Is it genetic or learned behaviour? What have freewill and choice got to do with it? How do you feel as a practitioner – positive? manipulated? impotent? Should we cope with our own workload or sub-specialise? How to we judge people? What has demography got to do with it?

- Training is available from RCGP

- Harm reduction is important
- Think about social/physical/psychological aspects
- Think physical prevention (primary, seconday and tertiary)

ENT and facial problems (statement 15.4)

You need to know:

- There's a lot of it about
- About deafness and the law – access to health care (Disability Discrimination Act 1995)
- About early diagnosis of head and neck cancers
- About detection/screening for deafness in children
- About referral guidelines

Did you know that some children who appear to be hard of hearing can distinguish the word "chocolate" whispered from behind them at surprisingly great distances?

Some apparently hard of hearing old people have a similar response to phrases such as "hot water bottle", "small sherry", "lottery win" and "Carol Vorderman's getting her kit off".

Skin problems (statement 15.10)

You need to know about

- urgent referral guidelines (skin cancers)
- the psychological effects of disfiguring skin problems (eg. acne in adolescents)
- skin health promotion/education (eg. the dangers of sunbathing and sunbeds)
- management of chronic skin conditions

If you don't do a dermatology job, consider going to come dermatology clinics. However, you will see a lot of skin problems in your general practice post, because there's a lot of problem skin about, as I discovered when I looked in the mirror just now.

Appendix: A Game

You might think that all the clever ideas and concepts in the GP Curriculum are new. Alternatively, you might not think they're all that clever, but still believe them to be new. Well, that's where you're wrong. Below is a list of key words and phrases, followed by some quotes from the *Meditations* of the famous Roman Emperor and Stoic philosopher Marcus Aurelius. The key words and phrases (as you can see) are all related to important ideas in the Curriculum. All you have to do is match the quotes to the key words and phrases, thus demonstrating that many of the ideas in the Curriculum go back at least as far as the Romans.

Key Words and Phrases:

    Duties
    Qualities
    Problem solving
    Community orientation
    Holism
    Self-awareness and attitudes
    Don't give up hope
    Duty and deontology
    Patient centredness
    Don't expect presents
    Value-based medicine
    Impermanence and death
    Evidence-based medicine

Retirement

Becoming an expert

Integrity

Ethics

Quotes:

… resolve firmly, to do what comes to hand with correct and natural dignity, and with humanity, independence, and justice.

Our mental powers should enable us to perceive the swiftness with which all things vanish away; their bodies in the world of space, and their remembrance in the world of time.

Things are determined by the view taken of them…

And even if man's years be prolonged, we must take into account that it is doubtful whether his mind will continue to retain its capacity for understanding of business or for the contemplative effort needed to apprehend things divine and human.

Treat with respect to power you have to form an opinion. By it alone can the helmsman within you avoid forming opinions that are at variance with nature and with the constitution and other reasonable being .

When an object presents itself to your perception, make a mental definition or at least an outline of it, so as to discern its essential character, to pierce beyond its separate attributes to distinct view of the naked whole …

Men seek for seclusion in the wilderness, by the seashore or in the mountains... such a fancy is wholly unworthy of a philosopher, since at any moment you choose you can retire within yourself. Avail yourself often, then of this retirement, and so continually renew yourself.

To do exclusively what reason ... shall suggest for the common weal; and to reconsider a decision if anyone present should correct you and convince you of an error of judgement. But such conviction must proceed from the assurance that justice, or the common good, or some other interests will be served. This must be the sole consideration; not the likelihood of pleasure or popularity

At day's first light have in readiness, against disinclination to leave your bed, the thought that " I am rising of the work of man"

If there is something good to be done or said, never renounce your right to it. Keep a straight course and follow your own nature.

You will never be a remarkable for quick- wittedness. Cultivate these, then,for they are wholly within your power; sincerity, for example, and dignity; industriousness, and sobriety. Avoid grumbling; be frugal, considerate, and frank; temperate in manner and speech.

As for truth, it is so veiled in obscurity that many reputable philosophers assert the impossibility of reaching any certain knowledge. All intellectual conclusions are fallible; for where is the infallible man?

The chief good of a rational being is fellowship with his neighbours ...

Look beneath the surface: never let a things intrinsic quality or worth escape you.

Work out, in action and inaction alike, the purpose of our natural constitutions.

Because the thing is difficult for you, do not therefore suppose it could be beyond mortal power. On the contrary, if anything is possible and proper for a man to do, assume that it must fall within your own capacity.

I do that which it is my duty to do.

Try to move men by persuasion; and act against their will if the principles of justice so direct. Turn the obstacle into an opportunity for exercise of some other virtue.

Accustom yourself to give careful attention to what others are saying, and try your best to enter into the mind of the speaker.

Is it possible for any useful thing to be achieved without change? Do you see, then, that change in yourself is of the same order, and no less necessary ...

Fix your thoughts closely on what has been said, and let your mind enter fully into what is being done, and into what is doing it.

When you have done a good action, and another has had the benefit of it, why crave for yet more in addition-applause of your kindness, or some favour in return-as the foolish do?

Enough if your present opinion be grounded in conviction, your present action grounded in unselfishness, and your present dispositions contented with whatever befalls you from without.

Facts stand wholly outside our gates; they are what they are, and no more; they know nothing about themselves, and they pass no judgement on themselves. What is it, then, that pronounces the judgement? Our own guide and ruler, Reason.

Your own mind, the mind of the universe, your neighbour's mind - be prompted to explore them all.

Waste no more time arguing what a good man should be. Be one.

What are the children of men, but as leaves that drop at the wind's breath?

What is your trade? Goodness. But how are you to make a success of it unless you have a philosopher's insight into the nature of the universe, and into the particular constitution of man.

Practice, even when success looks hopeless. The left-hand, inept in other respects for lack of practice, can grasp the rains more firmly than the right, because here it has had practice.

In the management of your principles, take example by the pugilist, not the swordsman. One puts down his blade and has to pick it up again; the other is never without his hand, and so needs only to clinch it.

Does the lantern's flame shine with undimmed brilliance until it is quenched, yet shall truth, wisdom, and justice die within you before you yourself are extinguished?

If it is not the right thing to do, never do it; if it is not the truth, never say it. Keep your impulses in hand.

Always look at the whole of the thing. Find what it is that makes its impression on you....

"But I have played no more than three of the five acts." Just so; in your drama of life, three acts are all the play. Pass on your way, then, with a smiling face ...

Here is a blank page for you to fill up in your own way. Why not draw a picture of your Trainer's smiling face?

Here is another blank page for you to fill in however you choose. Why not write a poem about some heroic deed you have seen your Trainer perform?